2018
室内设计模型集成
INTERIOR DESIGN MODEL INTEGRATION

公共空间
PUBLIC SPACE

叶斌 叶猛 著

海峡出版发行集团 | 福建科学技术出版社

作者简介 | AUTHOR PROFILE

叶 斌 Ye Bin

高级建筑师
国广一叶装饰机构首席设计师
福建农林大学兼职教授
南京工业大学建筑系建筑学学士
北京大学EMBA
中国室内设计学会理事
中国建筑装饰协会理事

Senior Architect
Chief Architect of Guoguangyiye Decoration Group
Adjunct Professor of Fujian Agriculture and Forestry University
B. Arch from Nanjing Industry University
EMBA from Beijing University
Association Member of China Institute of Interior Design
Councilor Member of China Building Decoration Association

荣誉

当选2013～2015年度福建省最具影响力设计师（排名第一）
荣获"中国室内设计杰出成就奖"
当选2009"金羊奖"中国十大室内设计师
当选中国建筑装饰行业建国60年百名功勋人物
当选1989~2009中国杰出室内设计师
当选1997~2007中国家装十年最具影响力精英领袖
当选1989~2004全国百位优秀室内建筑师
当选2004年度中国室内设计师十大封面人物

著作

1. 《2017室内设计模型库》系列（5册）
2. 《经典家居设计》系列（4册）
3. 《2016家居设计模型库》系列（4册）
4. 《新家居装修与软装设计》系列（4册）
5. 《2016公共空间模型库》
6. 《2015室内设计模型集成》系列（5册）
7. 《名师家装新图典》系列（3册）
8. 《2014空间模型集成》系列（5册）
9. 《2013家居空间模型集成》系列（3册）
10. 《2013公共空间模型集成》系列（2册）
11. 《2012室内设计模型集成》系列（5册）
12. 《2011公共空间模型》
13. 《2011家居空间模型》系列（3册）
14. 《2010公共空间模型》系列（2册）
15. 《2010家居空间模型》系列（3册）
16. 《2009室内设计模型》系列（5册）
17. 《概念家居》、《概念空间》
18. 《家居装饰·平面设计概念集成》
19. 《国广一叶室内设计模型库》系列
20. 《室内设计立面构图艺术》系列
21. 《国广一叶室内设计模型库构成元素》（1、2）
22. 《国广一叶室内设计》
23. 《国广一叶室内设计模型库·公建装饰》
24. 《国广一叶室内设计模型库·家居装饰》（1、2、3）
25. 《建筑外观细部图典》
26. 《装饰设计空间艺术·公共建筑装饰》
27. 《装饰设计空间艺术·家居装饰》（1、2、3）
28. 《室内设计图典》（1、2、3）

获奖设计作品

长乐电力大楼	2015～2016年度中国建筑工程装饰奖（公共建筑装饰设计类）
叶禅赋	2016第十一届中国国际室内设计双年展金奖
FORUS	2016第十一届中国国际室内设计双年展金奖
Lee House	2016第十一届中国国际室内设计双年展金奖
静·念	2016第十一届中国国际室内设计双年展银奖
仕林东湖	2016第十一届中国国际室内设计双年展银奖
《白说》	2016第十一届中国国际室内设计双年展银奖
"一扇窗，漫一室"	2016第十一届中国国际室内设计双年展银奖
坐卧之间	2016第十一届中国国际室内设计双年展铜奖
亦黑亦白	2016第十一届中国国际室内设计双年展铜奖
世欧澜山	2016第十一届中国国际室内设计双年展铜奖
秋风词	2016第十一届中国国际室内设计双年展铜奖
少即是多	2016第十一届中国国际室内设计双年展铜奖
溪山温泉度假酒店（实例）	2014年第十届中国国际室内设计双年展金奖
正兴养老社区体验中心	2014年第十届中国国际室内设计双年展银奖
中联大厦办公楼	2014年第十届中国国际室内设计双年展铜奖
尊贵彰显富丽	2014年第十届中国国际室内设计双年展铜奖
永福设计研发中心	2014年度全国建筑工程装饰奖（公共建筑装饰设计类）
宇洋中央金座	2013年第十六届中国室内设计大奖赛铜奖
宁德上东曼哈顿销售楼部	2013年第四届中国国际空间环境艺术大赛（筑巢奖）优秀奖
福建洲际酒店	2012年首届亚太金艺奖酒店设计大赛金奖
瑞莱春堂	2012年第四届"照明周刊杯"照明应用设计大赛金奖
前线共和广告	2012年第十五届中国室内设计大奖赛金奖
前线共和广告	2012年第九届中国室内设计双年展金奖
阳光理想城	2012年第九届中国室内设计双年展银奖
福州情·暴春园	2012年第九届中国室内设计双年展银奖
宁化世界客属文化交流中心	2012年第九届中国室内设计双年展铜奖
映·像	2012年第二十届业太室内设计大奖赛铜奖
名城港湾157#103	2012年第三届中国国际空间环境艺术设计大赛（筑巢奖）优秀奖
一信（福建）投资	2011年第十四届中国室内设计大奖赛金奖
福建科大永和医疗机构	2011年中国最成功设计大赛最成功设计奖
素丽娅泰SPA	2010年第八届中国室内设计双年展金奖
摩卡小镇销售楼中心	2010年第八届中国室内设计双年展银奖
大洋鹭洲	2010年第八届中国室内设计双年展铜奖
素丽娅泰SPA	2010年亚太室内设计双年展大奖赛商业空间设计银奖
繁都魅影	2010年亚太室内设计双年展大奖赛住宅空间设计银奖
繁都魅影	2010年业太室内设计大奖赛铜奖
中央美苑	2010海峡两岸室内设计大赛金奖
繁都魅影	2010海峡两岸室内设计大赛金奖
光·盒中盒	2010海峡两岸室内设计大赛金奖
皇帝洞书院	2009年"尚高杯"中国室内设计大奖赛二等奖
北湖皇帝洞景区会所	2008年第七届中国室内设计双年展金奖
大家会馆（实例）	2006年第六届中国室内设计双年展金奖
书香大第销售中心	2006年第六届中国室内设计双年展金奖

另120多项设计作品荣获福建省室内设计大奖赛一等奖

叶 猛 Ye Meng

国广一叶装饰机构副总设计师
国家一级注册建筑师
国家一级注册建造师
中国建筑学会室内分会会员
福建工程学院建筑与规划系讲师
福州大学建筑系学士
中南大学土建学院建筑学硕士

Deputy Chief Architect of Guoguangyiye Decoration Group
First-Class Registered Architect (PRC)
Registered Constructor (PRC)
Member of Institute of Interior Design of Architectural Society of China
Lecturer of Architecture and Planning Dept., Fujian University of Technology
B. Arch from Fuzhou University
M. Arch from Central South University

获奖设计作品

仕林东湖	2016第十一届中国国际室内设计双年展银奖
融信大卫城—禅韵	2016福建省室内设计大赛居室空间类金奖
东方韵	2015中南地区国际空间环境艺术设计大赛方案设计空间铜奖
雅韵·世欧澜山	2015中南地区国际空间环境艺术设计大赛住宅空间优秀奖
风尚	2015年度国际空间设计大奖·艾特奖 最佳公寓设计入围奖
融信大卫城	2014年第十届中国国际室内设计双年展优秀奖
三盛滨江公园	2014年第五届中国国际空间环境艺术大赛（筑巢奖）提名奖
名城港湾	2014年第五届中国国际空间环境艺术大赛（筑巢奖）优秀创意奖
融侨外滩	2014年第五届中国国际空间环境艺术大赛（筑巢奖）优秀创意奖
鳌峰洲小区—19A	2013年第四届中国国际空间环境艺术设计大赛（筑巢奖）优秀奖
阳光理想城	2012年第九届中国国际室内设计双年展金奖
大洋鹭洲	2010年第八届中国室内设计双年展铜奖
繁都魅影	2010年亚洲室内设计大奖赛铜奖
福建工程学院建筑系新馆	2009年中国室内环境艺术设计大赛一等奖
福建工程学院建筑系新馆	2009年福建室内与环境设计大奖赛公建工程类最高奖
文化主题酒店	2008年福建省第六届室内与环境设计大赛一等奖
点房财富中心	2007年"华耐杯"中国室内设计大奖赛二等奖
大家会馆（实例）	2006年第六届中国室内设计双年展金奖

另出版《建筑外观细部图典》、《室内设计图像模型》等著作数十种

前言 / PREFACE

国广一叶装饰机构作为"全国最具影响力室内设计机构"（中国建筑学会室内设计分会颁发）、2017年第二十届中国室内设计大奖赛"最佳设计企业"（中国建筑学会室内设计分会颁发）、"2016年度中国建筑装饰杰出住宅空间设计机构"（中国建筑装饰协会颁发）、"2015年度中国建筑装饰设计机构50强企业"（中国建筑装饰协会颁发）、"2013住宅装饰装修行业最佳设计机构"（中国建筑装饰协会颁发）、2013年度全国住宅装饰装修行业百强企业（中国建筑装饰协会颁发）、"2012～2013年度全国室内装饰优秀设计机构"（中国室内装饰协会颁发）、"2012年中国十大品牌酒店设计机构"（中外酒店论证颁发）、"2013中国住宅装饰装修行业最佳设计机构"（中国建筑装饰协会颁发）、"1989～2009年全国十大室内设计企业"（中国建筑协会室内设计分会颁发）、"1988～2008年中国室内设计十佳设计机构"（中国室内装饰协会颁发）、"福建省著名商标"、"福建省建筑装饰装修行业龙头企业"（福建省人民政府闽政文〔2014〕26号颁发），"福建省建筑装饰行业协会会长单位"，荣获国际、国家及省市级设计大奖上千项。

国广一叶装饰机构首席设计师叶斌荣获"中国室内设计杰出成就奖"、两次荣获"中国十大室内设计师"称号；叶猛被评为"1989～2009年中国优秀设计师"、"福建十大杰出（住宅空间）设计师"称号；另外，38名设计师被评为中国装饰设计行业优秀设计师，121名设计师分别被评为福建省优秀设计师、福州市优秀设计师，101名在职设计师分别荣获历届全国、福建省、福州市室内设计一等奖……以上这些荣誉的获得和国广一叶装饰机构自身的水准有关。国广一叶装饰机构拥有大批量高水准的室内设计专业效果图，这些效果图将设计师的设计意图淋漓尽致地表现出来。

自2004年至今，国广一叶装饰机构在福建科学技术出版社已陆续出版了20套模型系列图书，共54本模型系列图书，一直受到广大读者的支持与厚爱。为了不辜负广大读者的期望，我们继续推出《2018室内设计模型集成》系列图书。这系列图书汇集了国广一叶装饰机构2017～2018年制作的1500多个风格各异、手法时尚的室内设计效果图及其对应的3ds Max场景模型文件，可用来被作为读者做室内设计时的有益参考。

本书配套光盘的内容包含效果图原始3ds Max模型和使用到的所有贴图文件。由于3ds Max软件不断升级，此次的模型我们采用3ds Max2014版本制作。模型按图片顺序编排，易于查阅和调用。只有能对模型进一步调整才能体现其价值和生命力，因此提供的3ds Max模型是真正有价值、可随时提取调整用的部分。必须说明的是，书中收录的效果图均为原始模型经过lightscape渲染和photoshop后期处理过的成图，是为读者了解后处理效果提供直观准确的参考，与3ds Max直接渲染的效果有一定区别。

著 者
2018年2月

As "most influential Interior Design Companies in China" (honored by China Building Association Interior Design Chapter), Guoguangyiye Decoration Group have acquired thousands of international, national and provincial design awards, such as "the best design company (honored by Twentieth China Interior Design Grand Prix in 2017)" "China Building Decorationg Outstangding Residence Space Design Institutions in 2016 (honored by China Building Decoration Association, CBDA) " "Top 50 architectural decoration company in China(2015, honored by China Building Decoration Association, CBDA)", "the best design institutions of residential ornament industry in 2013 (honored by China Building Decoration Association, CBDA)", "the Top 100 enterprises of Chinese residential ornament industry in 2013(honored by China Building Decoration Association, CBDA)" "Outstanding Interior Design Companies in China(2012-2013, honored by China National Interior Decoration Association, CIDA)", "The Best Interior Decoration Association of Chinese Home Decoration(2013, honored by CBDA)", "Top 10 Interior Design Companies in China (1989~2009, honored by China Building Association Interior Design Chapter)", "Top 10 China Interior Design Institutions (1988~2008, issued by China Interior Decoration Institutions)", "Well-Known Brand of Fujian", "the leading enterprises of architectural ornament industry in Fujian province (issued by the people's Government of Fujian Province〔2014〕No. 26)" and "the president company of Architectural Ornament Industry Association of Fujian province".

The chief architect Mr. Bin Ye has wined the award of "Distinguished Achievement Award of Chinese Interior Design", and awarded twice "China Top 10 Interior Design Architect". Mr. Meng Ye was awarded "Outstanding Architect of China (1989~2009)"."Fujian Top 10 Outstanding Residential space Architect" ..Besides, 38 architects have awarded as "Excellent Architect of China Decoration Design Industry", 121 architects have awarded as "Excellent Architect of Fujian province/Fuzhou",,and 101 architects have won top prize of national, Fujian provincial or Fuzhou.

Naturally these achievements have been accomplished because of the high level interior designs of Guoguangyiye, but obviously cannot be attained without high level professional effect drawing that presents the design intent of architects incisively and vividly. Therefore as a product of the collective efforts of architect and graphic designer, it is closely related to the success of project design.

Since 2004, Guoguangyiye has published 20 offsets, a total of 54 books ,on design model database with Fujian Science and Technology Press and all of them have gained wide popularity by their richness and practicality. Therefore, this year we will continue to publish 2018 Interior Design Model Integration. This new series consists of over 1500 chic 3ds Max scenario models of various style interior designs created by Guoguangyiye Decoration Group during 2017~2018. Being a model database, they could also be used as beneficial references for interior design.

The enclosed DVD contains original 3ds Max models of decoration effect drawings and all the map files used in order to create them. Due to the continuous upgrading of 3ds Max software, version 2014 was adopted in the drawing of these models which are arranged in the order of the pictures to make them easily lookup and call. Since as only models that can be further adjusted are valuable, the 3ds Max moulds provided are all of true value and readily available. It should be noted that, all the effect drawings in the books are pictures rendered by lightscape and dealt with by Photoshop, to give an intuitive and precise reference for readers on the after effects which are different from those rendered directly by 3ds Max.

February 2018

CONTENTS / 目录

办公空间 OFFICE SPACE — 005

商业空间 COMMERCIAL SPACE — 063

学校空间 SCHOOL SPACE — 113

房产空间 REAL ESTATE SPACE — 131

OFFICE SPACE

办公空间

001
办公区走道
Open office area walkway

002
董事长办公室
Chairman's Office

003
办公区走道
Open office area walkway

004
办公区
Office area

005
办公区走道
Open office area walkway

006 会议室 Meeting room

007 总经理办公室 General Manager's Office

008 办公区
Office area

009 服务区
Service area

/ 办公空间 **013**
OFFICE SPACE

010
会议室
Meeting room

011
多功能厅
Multi-function hall

012
大堂
Lobby

013
走道
Walkway

014
办公区走道
Open office area walkway

015 多功能厅 Multiple-function hall

016 办公室 Office

017
办公室
Office

018
总经理办公室
General Manager's Office

019 洗手间 Washroom

020 门厅 Foyer

021 会议室 Meeting room

022 办公室 Office

023 办公室 Office

024
服务大厅
Service hall

026
展厅
Exhibition hall

025
会议室
Meeting room

027
展厅
Exhibition hall

028 办公室 Office

029 电梯厅 Elevator hall

030 接待室 Reception room

031
接待厅
Reception hall

会议室
Meeting room

会议室
Meeting room

034
总经理办公室
General Manager's Office

035
会议室
Meeting room

036
办公区
Office area

037
办公区
Office area

038
会议室
Meeting room

039
会议室
Meeting room

040
走道
Walkway

041 门厅 Foyer

042 办公区 Office area

043 会议室 Meeting room

044 办公室 Office

047
展厅
Exhibition hall

046
展厅
Exhibition hall

049
洗手间
Washroom

048
展厅
Exhibition hall

050
洗手间
Washroom

051 办公室 Office

052 办公室 Office

053 会议厅 Conference room

054 会议厅 Conference room

055 会议厅 Conference room

056
总经理办公室
General Manager's Office

057
接待室
Reception room

058 会议厅 Conference room

059 服务大厅 Service hall

/ 办公空间　OFFICE SPACE　037

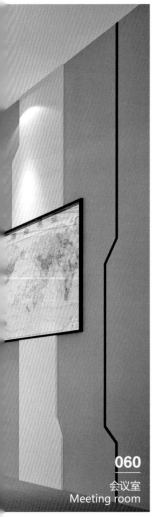

060
会议室
Meeting room

061
会议室
Meeting room

062
入口
Entrance

063
操作室
Control room

064
办公室
Office

065 多功能厅 Multiple-function hall

066 董事长办公室 Chairman's Office

067 门厅 Foyer

070
会议室
Meeting room

/ 办公空间　041
OFFICE SPACE

072
总经理办公室
General Manager's Office

069
董事长办公室
Chairman's Office

071
电梯厅
Elevator hall

073
董事长办公室
Chairman's Office

/ 办公空间
OFFICE SPACE 043

074
大堂
Lobby

075
会议室
Meeting room

076
走道
Walkway

077
走道
Walkway

078
走道
Walkway

079 办公区 Office area

080 办公区 Office area

081
会议室
Meeting room

082
会议室
Meeting room

083 会议室 Meeting room

084 总经理办公室 General Manager's Office

085
服务大厅
Service hall

086
服务大厅
Service hall

087 会议厅 Conference room

088 会议室 Meeting room

089 会议室 Meeting room

090 办公区
Office area

091 办公室
Office

092 会议室 Meeting room

093 办公区 Office area

094 办公区 Office area

096 董事长办公室 Chairman's Office

097 董事长办公室 Chairman's Office

098
电梯厅
Elevator hall

099
会议室
Meeting room

100 门厅 Foyer

101 办公区 Office area

102 办公区 Office area

104
总经理办公室
General Manager's Office

105
总经理办公室
General Manager's Office

106
董事长办公室
Chairman's Office

107
办公室
Office

办公室 Office

总经理办公室 General Manager's Office

110
门厅
Foyer

111
门厅
Foyer

112 总经理办公室
General Manager's Office

113 门头
Door header

114 电梯厅
Elevator hall

115
办公区
Office area

116
前台
Information desk

117
走道
Walkway

118 前台 Information desk

119 前台 Information desk

COMMERCIAL SPACE

商业空间

120
银行外观
Bank appearance

121
银行外观
Bank appearance

122
营业厅
Business hall

123
营业厅
Business hall

124
营业厅
Business hall

125
展厅
Exhibition hall

126
展厅
Exhibition hall

127 展厅
Exhibition hall

128 展厅
Exhibition hall

129 走道 Walkway

130 餐厅走道 Restaurant walkway

阅览室
Reading room

132 展厅
Exhibition hall

133 银行外观
Bank appearance

134
银行外观
Bank appearance

135
银行外观
Bank appearance

136 营业厅 Business hall

137 营业厅 Business hall

138
营业厅
Business hall

139
营业厅
Business hall

140
宴会厅
Banquet hall

141 新华书店吧台
Xinhua Bookstore bar

142 新华书店
Xinhua Bookstore

143
新华书店
Xinhua Bookstore

144
新华书店
Xinhua Bookstore

145
机场贵宾厅
Airport VIP hall

146
机场贵宾厅
Airport VIP hall

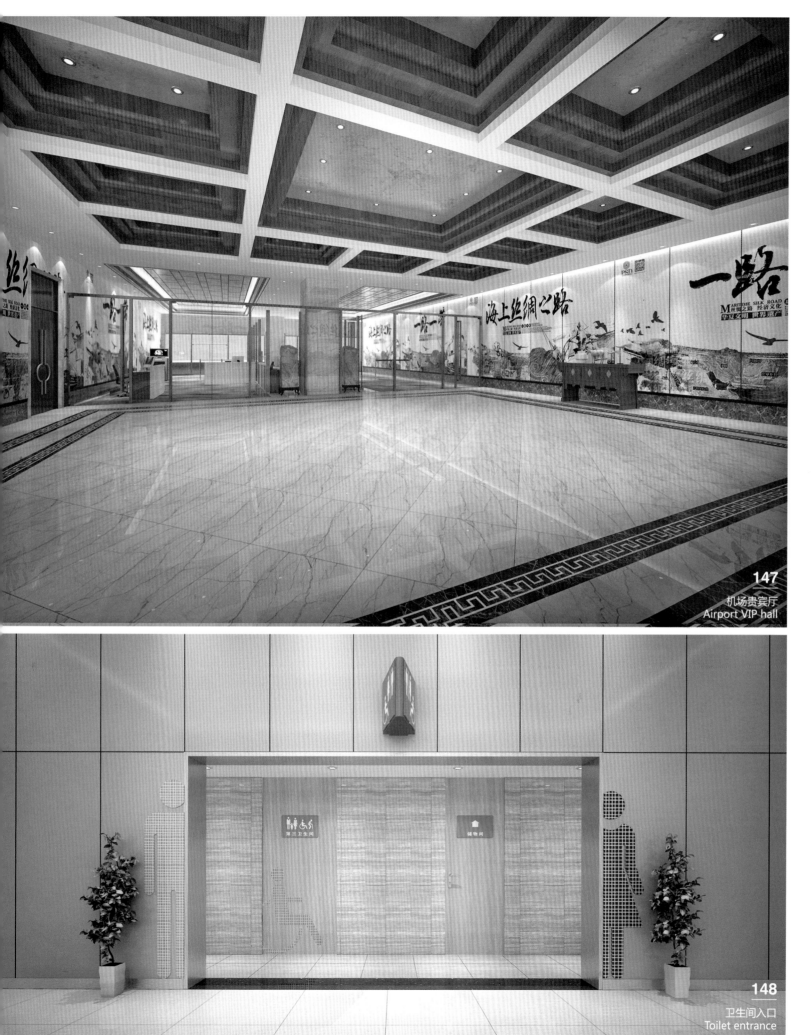

147
机场贵宾厅
Airport VIP hall

148
卫生间入口
Toilet entrance

149
服务大厅
Service hall

150
机场大厅
Airport hall

151 营业厅 Business hall

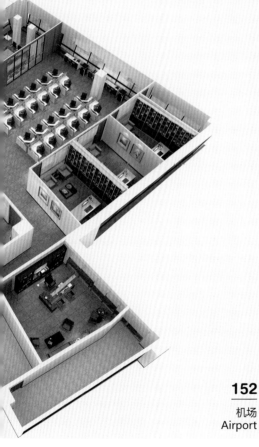

152 机场 Airport

153 机场快线站 Airport Express Station

/ 商业空间　COMMERCIAL SPACE　083

155
营业厅
Business hall

154
营业厅
Business hall

155
营业厅
Business hall

156
旅游集散中心
Tourist distributing center

157
旅游集散中心
Tourist distributing center

158
旅游集散中心
Tourist distributing center

159 新华书店
Xinhua Bookstore

160 新华书店
Xinhua Bookstore

161 新华书店 Xinhua Bookstore

162 新华书店 Xinhua Bookstore

163 酒楼过道 Restaurant aisle

164 会所门厅
Club hall

165 银行大堂
Bank lobby

茶艺室
Tea room

167 银行外观 Bank appearance

168 营业厅 Business hall

169 营业厅 Business hall

170 营业厅 Business hall

171 营业厅 Business hall

172
新华书店
Xinhua Bookstore

173
新华书店
Xinhua Bookstore

174
新华书店
Xinhua Bookstore

175
新华书店
Xinhua Bookstore

176
营业厅
Business hall

177
银行接待厅
Bank reception hall

/ 商业空间　093
COMMERCIAL SPACE

178
银行接待厅
Bank reception hall

179
餐厅包间
Dining room

180
机场贵宾厅
Airport VIP hall

181
银行门厅
Bank lobby

182
银行贵宾厅
Bank VIP hall

183
银行贵宾厅
Bank VIP hall

184
银行接待厅
Bank reception hall

/ 商业空间
COMMERCIAL SPACE

187
银行门厅
Bank lobby

185
银行接待厅
Bank reception hall

188
银行贵宾室
Bank VIP room

186
餐厅包间
Dining room

189
展厅
Exhibition hall

191
书店楼梯
Bookstore staircase

/ 商业空间 **101**
COMMERCIAL SPACE

193
饼屋
Cake house

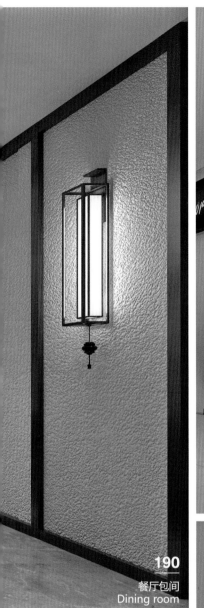

190
餐厅包间
Dining room

192
健身房
Fitness centre

194
饼屋
Cake house

195
银行入口
Bank entrance

196
营业厅
Business hall

197
卫生间
Washroom

198
营业厅
Business hall

199
营业厅
Business hall

200
娱乐室
playroom

201
书店楼梯
Bookstore staircase

/ 商业空间
COMMERCIAL SPACE
105

202 营业厅 Business hall

203 门厅 Foyer

204 营业厅 Business hall

205
酒吧楼梯
Bar staircase

206
火锅店
Hotpot restaurant

207
火锅店
Hotpot restaurant

208
火锅店
Hotpot restaurant

209
酒吧
Bar

210
酒吧
Bar

211
餐厅包间
Dining room

/ 商业空间
COMMERCIAL SPACE

212
宴会厅
Banquet hall

213
门头
Door header

214
门头
Door header

215
服务大厅
Service hall

216
营业厅
Business hall

SCHOOL SPACE

学校空间

/ 学校空间 115
SCHOOL SPACE

217
阅览区
Reading room

218
阅览区
Reading room

219
门厅
Foyer

220
门厅
Foyer

221
阅览区
Reading room

222
阅览区
Reading room

/学校空间
SCHOOL SPACE 117

223
阅览区
Reading room

224
教室
Classroom

报告厅
lecture hall

226
幼儿园
Kindergarten

227
幼儿园
Kindergarten

228 教室 Classroom

229 教室 Classroom

230 报告厅 lecture hall

231
图书室
Library

232
图书室
Library

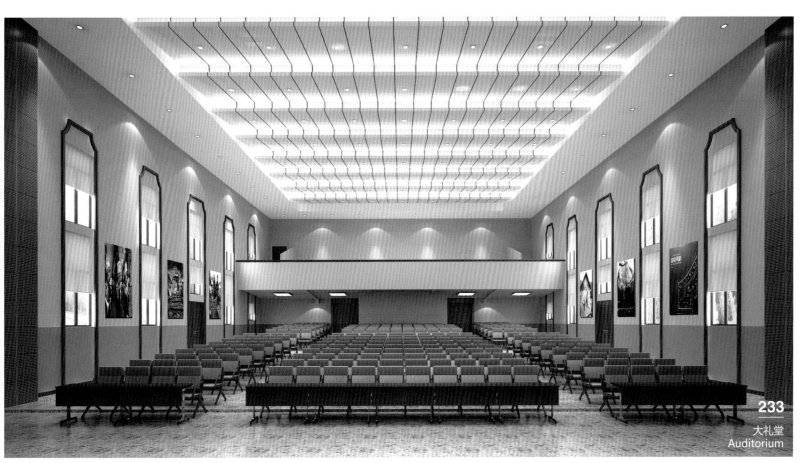

233 大礼堂 Auditorium

234 阅览区 Reading room

235
幼儿园
Kindergarten

236
幼儿园
Kindergarten

237
幼儿园
Kindergarten

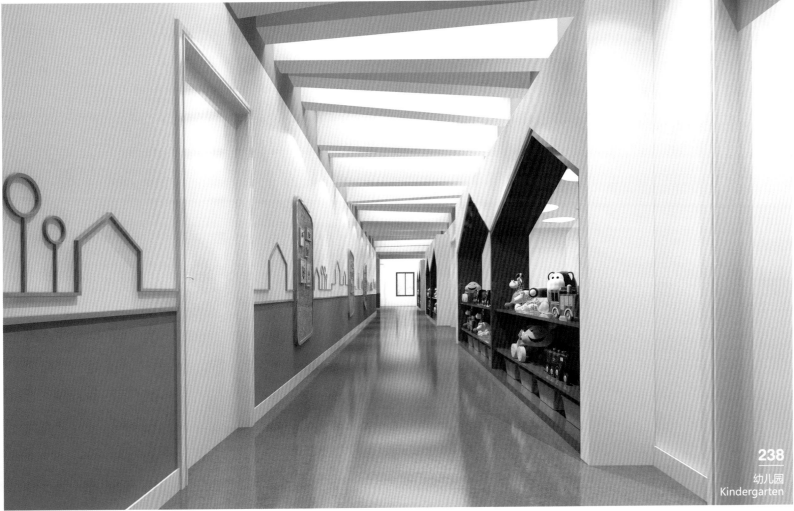

238
幼儿园
Kindergarten

239 展厅 Exhibition hall

240 展厅 Exhibition hall

241
图书馆
Library

242
培训室
Training room

243 培训室 Training room

244 教室 Classroom

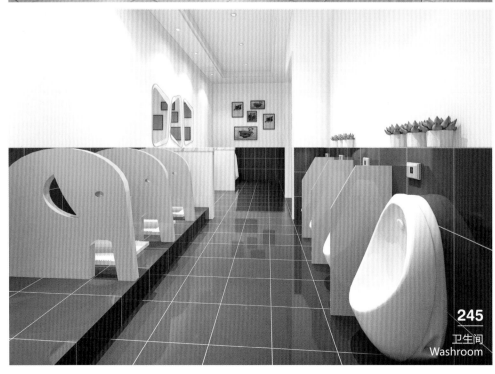

245 卫生间 Washroom

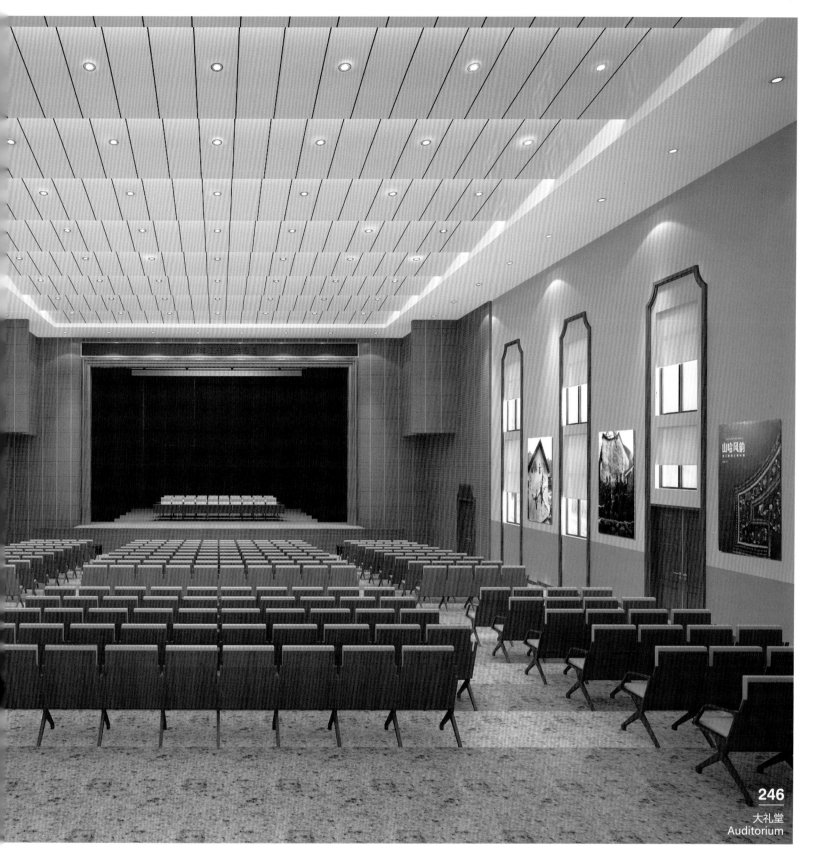

246
大礼堂
Auditorium

247
科普室
Popular science room

248
学校图书馆
School library

REAL ESTATE SPACE

房产空间

单元样板房
Unit show flat

250 单元样板房
Unit show flat

251 单元样板房
Unit show flat

252
单元样板房
Unit show flat

253
单元样板房
Unit show flat

254
电梯厅
Elevator hall

255
电梯厅
Elevator hall

256 洽谈区
Negotiate area

257 过道
Passageway

258 电梯厅
Elevator hall

259
售楼中心
Sales center

260
售楼中心花园
Sales center garden

261
单元样板房
Unit show flat

262
活动室
Activity room

/ 房产空间　　141
REAL ESTATE SPACE

263
接待室
Reception room

264
接待厅
Reception hall

265
单元样板房
Unit show flat

266
单元样板房
Unit show flat

267
单元样板房
Unit show flat

268
单元样板房
Unit show flat

/ 房产空间
REAL ESTATE SPACE   143

269
单元样板房
Unit show flat

270
电梯厅
Elevator hall

271
电梯厅
Elevator hall

272
别墅样板房
Villa show flat

273
别墅样板房
Villa show flat

274
别墅样板房
Villa show flat

275
单元样板房
Unit show flat

276 过道 Passageway

277 单元样板房 Unit show flat

278 卫生间 Washroom

/ 房产空间
REAL ESTATE SPACE 147

279
别墅样板房
Villa show flat

280
别墅样板房
Villa show flat

281
别墅样板房
Villa show flat

282
别墅样板房
Villa show flat

283
单元样板房
Unit show flat

284
别墅样板房
Villa show flat

285 过道 Passageway

286 过道 Passageway

287 SPA 房 SPA room

接待室
Reception room

289
入户门厅
Entrance hall

290
入户门厅
Entrance hall

291 别墅样板房
Villa show flat

292 售楼大厅
Sales hall

293
别墅样板房
Villa show flat

294
别墅样板房
Villa show flat

295
单元样板房
Unit show flat

296
单元样板房
Unit show flat

297
单元样板房
Unit show flat

299
别墅样板房
Villa show flat

300
别墅样板房
Villa show flat

301
别墅样板房
Villa show flat

302
别墅样板房
Villa show flat

/ 房产空间　**161**
REAL ESTATE SPACE

303
别墅样板房
Villa show flat

306
单元样板房
Unit show flat

305
别墅样板房
Villa show flat

307
单元样板房
Unit show flat

308
单元样板房
Unit show flat

309
单元样板房
Unit show flat

310
别墅样板房
Villa show flat

311
别墅样板房
Villa show flat

312 单元样板房 Unit show flat

313 单元样板房 Unit show flat

314 单元样板房 Unit show flat

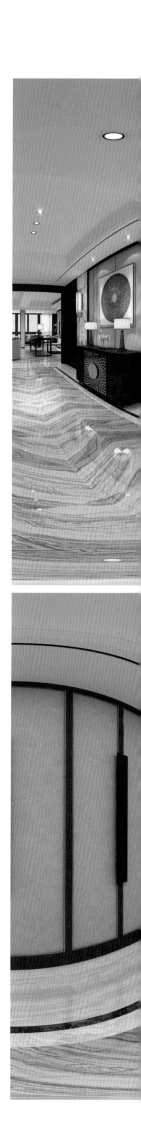

315
别墅样板房
Villa show flat

316
别墅样板房
Villa show flat

单元样板房
Unit show flat

房产空间 167
REAL ESTATE SPACE

320
单元样板房
Unit show flat

317
单元样板房
Unit show flat

321
别墅样板房
Villa show flat

319
单元样板房
Unit show flat

322
别墅样板房
Villa show flat

323
别墅样板房
Villa show flat

324
别墅样板房
Villa show flat

325
单元样板房
Unit show flat

326
单元样板房
Unit show flat

327
别墅样板房
Villa show flat

328
别墅样板房
Villa show flat

330 别墅样板房
Villa show flat

331 别墅样板房
Villa show flat

332 别墅样板房
Villa show flat

/ 房产空间
REAL ESTATE SPACE 173

333
别墅样板房
Villa show flat

334
单元样板房
Unit show flat

335
单元样板房
Unit show flat

336
单元样板房
Unit show flat

337
单元样板房
Unit show flat

338
单元样板房
Unit show flat

图书在版编目（CIP）数据

2018室内设计模型集成. 公共空间/叶斌，叶猛著. — 福州：福建科学技术出版社，2018.3
ISBN 978-7-5335-5556-6

Ⅰ.①2… Ⅱ.①叶…②叶… Ⅲ.①公共建筑－室内装饰设计－图集 Ⅳ.①TU238.2-64

中国版本图书馆CIP数据核字（2018）第030679号

书　　名	2018室内设计模型集成　公共空间
著　　者	叶斌　叶猛
出版发行	海峡出版发行集团
	福建科学技术出版社
社　　址	福州市东水路76号（邮编350001）
网　　址	www.fjstp.com
经　　销	福建新华发行（集团）有限责任公司
印　　刷	恒美印务（广州）有限公司
开　　本	635毫米×965毫米　1/8
印　　张	22
图　　文	176码
版　　次	2018年3月第1版
印　　次	2018年3月第1次印刷
书　　号	ISBN 978-7-5335-5556-6
定　　价	288.00元

书中如有印装质量问题，可直接向本社调换